Section 3

Organisms exchange substances with their environment

Surface area to volume ratio

Smaller organisms have a large surface area to volume ratio, which means that no cell is far from the body's surface. This makes it easier for the organism to exchange substances with its environment. Larger organisms have a smaller surface area to volume ratio,

meaning that they have adaptations for exchange of substances with their environment. This may involve:
- changes in body shape
- development of the body's systems for the exchange of substances

1 **a** Complete the table which compares the surface areas of various cubes to their volumes. (AO2)

2 marks

Length of one side of cube/mm	Surface area of one face of cube/mm²	Surface area of whole cube/mm²	Volume of cube/mm³	Surface area to volume ratio
1	1	6	1	6 : 1
2				
3				
4				
5				

b Describe the trend shown in the table. (AO2)

1 mark

....

....

2 The arctic fox has small ears, short legs relative to its body size and a short snout. The fennec fox is smaller than the arctic fox and lives in tropical areas. It has extremely large thin ears, a long snout and long thin legs relative to the slender body. Explain how these features adapt the different foxes to their environments. (AO2)

3 marks

....

....

....

....

....

....

3

3 The graph shows the metabolic rate of organisms with different body masses.

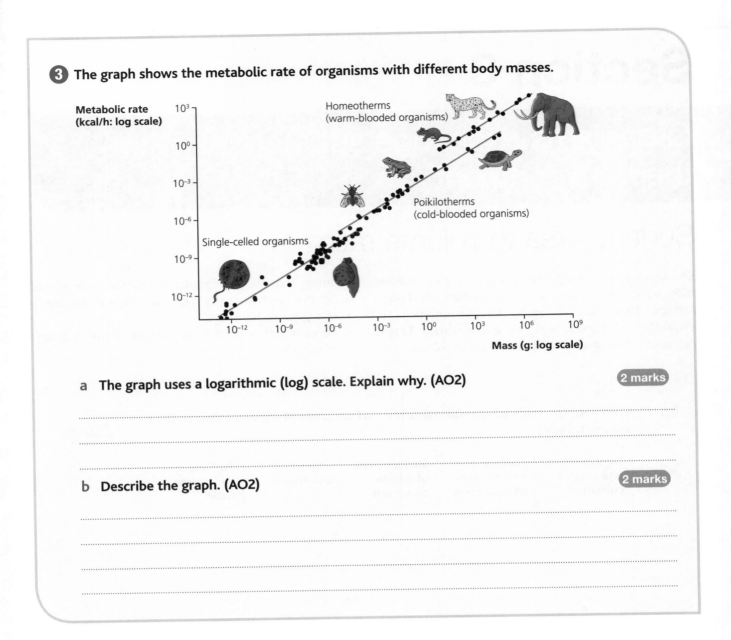

a   The graph uses a logarithmic (log) scale. Explain why. (AO2)   `2 marks`

....

....

....

b   Describe the graph. (AO2)   `2 marks`

....

....

....

....

....

Gas exchange

Single-celled organisms can exchange gases without having a specialised gas exchange system. Larger organisms have adaptations for gas exchange, such as the tracheal system in insects, gills in fish and lungs in mammals. Gas exchange in the leaves of dicotyledonous plants involves stomata allowing air into air spaces in the leaves so that most of the surface of the mesophyll cells is exposed to air. Organisms living on land lose water as a result of these systems, so many organisms, including terrestrial insects and xerophytic plants, have adaptations to reduce water loss. The human gas exchange system is specialised for efficient gas exchange, and movement of the intercostal muscles and diaphragm bring about ventilation.

1 Explain why a single-celled organism can exchange gases without having a specialised gas exchange system. (AO1)   `2 marks`

....

....

....

....

2 a Describe how gas exchange occurs in a terrestrial insect. (AO1)   `2 marks`

....

....

....

b When insects are active, lactate builds up in muscle tissue. This leads to an increase in the rate of diffusion of oxygen into the muscle tissue. Suggest how. (AO2)   `3 marks`

....

....

....

....

....

3 The graph shows the partial pressure of oxygen and carbon dioxide in the tracheae of an insect and the movements of the spiracles.

a Use the information in the graph to explain *one* adaptation to reduce water loss in this insect. (AO2)   `2 marks`

....

....

....

....

b A student decided that a low concentration of oxygen in the tracheae is the stimulus that causes the spiracles to open. Is this correct? Use evidence from the graph to explain your answer. (AO2)   `2 marks`

....

....

....

....

4 Describe how the gills of a fish are adapted for efficient gas exchange. (AO1) 5 marks

5 A fetus obtains oxygen by diffusion from its mother's blood in the placenta. In the placenta, the fetal blood vessels run very close to the mother's blood vessels, but the blood flow is in opposite directions. Explain the advantage of this. (AO2) 2 marks

6 a What is the gas exchange surface in a dicotyledonous leaf? (AO1) 1 mark

   b How is a dicotyledonous leaf adapted for efficient gas exchange? (AO1) 2 marks

7 a What is a xerophyte? (AO1) 1 mark

   b Explain *six* features that may be found in a xerophyte. (AO1) 6 marks

....
....
....
....
....

8 **Complete the table to show the features of gas exchange surfaces. (AO1)**   12 marks

Gas exchange in	Features that provide...		
	a large surface area	a large concentration gradient	a short diffusion pathway
Tracheal system of an insect			
Gills of fish			
Dicotyledonous leaf			
Mammalian lungs			

9 Complete the table to describe what happens during ventilation in a mammal. (AO1) 6 marks

	Inspiration (inhalation)	Expiration (exhalation)
Diaphragm		
External intercostal muscles		
Internal intercostal muscles		
Volume of thorax		
Pressure inside thorax		
Flow of air		

10 The graph shows the relative risk of developing lung cancer according to the number of cigarettes smoked per day.

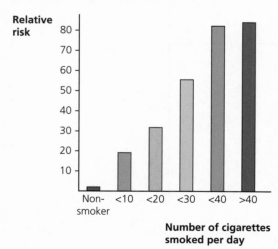

a   Describe the graph. (AO2)   2 marks

....

....

....

....

b   Do these data prove that smoking causes lung cancer? Explain your answer. (AO3)   2 marks

....

....

....

....

c   Suggest how the relative risk is calculated. (AO3)   2 marks

....

....

....

....

11 FEV1 is the volume of air that a person can force out of their lungs in 1 second.
A healthy person can expel at least 75% of the air in their lungs in the first second of
breathing out. Asthma is a condition in which the airways become narrower.

The table shows some data from two students of similar age, body size and sex.

Student	FEV1/dm³	Total volume of air breathed out/dm³	% of lung volume expelled in first second	Time taken to expel all air from lungs/s
Healthy	3.6	4	90	1.5
Asthma	1.8	3		2.5

a   Complete the table by calculating the % of lung volume expelled in the first
second by the student with asthma. (AO2)   1 mark

b   Explain why FEV1 is lower for the student with asthma. (AO2)   1 mark

....

....

Digestion and absorption

Digestion involves the action of enzymes in hydrolysing large molecules into smaller molecules that can be absorbed into the bloodstream. Starch is digested by amylase into maltose. Disaccharides are hydrolysed into monosaccharides by membrane-bound disaccharidases. Lipases digest lipids, whereas bile salts emulsify the lipid droplets, making digestion more efficient. Proteins are digested by endopeptidases, exopeptidases and dipeptidases in the membrane of the gut epithelial cells. The products of digestion are then absorbed into the cells lining the small intestine.

1 What is digestion? (AO1) `2 marks`

....

....

....

2 Name the type of reaction involved when large food molecules are digested. (AO1) `1 mark`

....

3 What is the difference between endopeptidases and exopeptidases? (AO1) `2 marks`

....

....

....

....

....

4 How do bile salts assist lipid digestion? (AO1) `2 marks`

....

....

....

....

....

5 Complete the table. (AO1) `8 marks`

Name of enzyme	Site of enzyme	Reaction catalysed
	Mouth	
	Stomach	
	Small intestine	Proteins to polypeptides
	Small intestine	Polypeptides to dipeptides
Amylase		
Lipase		
Maltase		
Dipeptidase		

6 What are micelles? Describe their importance in the absorption of lipids from the small intestine. (AO1) `3 marks`

....

....

....

....

....

....

....

7 Describe how amino acids are absorbed from the small intestine. (AO1)   5 marks

....

....

....

....

....

....

....

....

....

....

8 The diagram shows the co-transport of glucose into an epithelial cell from the small intestine.

Explain the effect that a lack of ATP would have on the absorption of glucose. (AO1)   4 marks

....

....

....

....

....

....

....

....

....

....

Exam-style questions

1 **a** Describe how oxygen reaches the muscle fibres of an insect. (AO1) 2 marks

....

....

....

....

b The giant dragonfly *Meganeura* lived 300 million years ago and had wings about 300 mm long. Modern insects are much smaller. In an investigation, scientists measured the maximum wing length of insects from different periods of evolutionary history and compared this to the oxygen concentration in the atmosphere at the time. The graph shows their results.

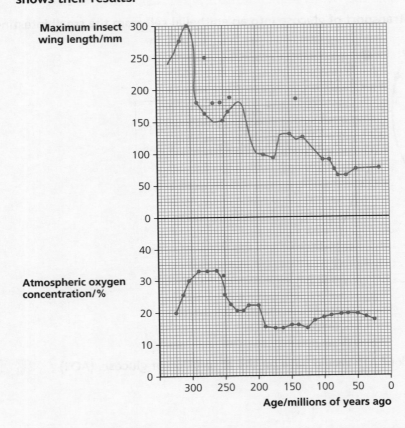

i How far do these results support the hypothesis that larger insects cannot survive in an atmosphere that is only 20% oxygen? (AO3) 3 marks

....

....

....

....

....

....

ii Use your knowledge of the insect gas exchange system to suggest how a higher oxygen concentration in the air might enable larger insects to survive. (AO2)

`3 marks`

....

....

....

....

....

....

....

....

Mass transport

Mass transport in animals

Haemoglobin is a protein with a quaternary structure. Slightly different types of haemoglobin are found in different organisms. Haemoglobin is present in red blood cells where it transports oxygen. The loading and unloading of oxygen by haemoglobin can be shown in an oxyhaemoglobin dissociation curve. Carbon dioxide affects the shape of the oxyhaemoglobin dissociation curve (the Bohr effect). Organisms living in different environments have different types of haemoglobin, which differ in their affinity for oxygen. The heart pumps blood around the blood vessels. Arteries, veins and capillaries have different structures that adapt them to their functions. Valves in the heart ensure a one-way flow of blood throughout the cardiac cycle. Tissue fluid forms at the arteriole end of capillaries, which supplies the body cells with glucose and oxygen.

1 The binding of oxygen to haemoglobin to form oxyhaemoglobin is said to be cooperative. Explain why. (AO1)

`2 marks`

....

....

....

....

2 Explain why haemoglobin is said to have a quaternary structure. (AO1)

`1 mark`

....

....

3 The graph shows the oxyhaemoglobin dissociation curve for human haemoglobin.

Percentage saturation of haemoglobin with oxygen

Partial pressure of oxygen/kPa

Use the graph to describe the features of haemoglobin that make it an efficient molecule to transport oxygen to different parts of the body. (AO2)   3 marks

....

....

....

....

....

....

4 Fetal haemoglobin has a dissociation curve to the left of the curve for adult haemoglobin. Explain the advantage of this. (AO2)   3 marks

....

....

....

....

....

....

5 The oxyhaemoglobin dissociation curve in question 3 is when there is a very low partial pressure of carbon dioxide.

a   Sketch a line on the graph to show the position of the oxyhaemoglobin dissociation curve when there is a higher partial pressure of carbon dioxide. (AO1)   1 mark

b   Explain the advantage of this. (AO1)   3 marks

....

....

....

....

....

....

6 The diagram shows the circulation of blood in a mammal.

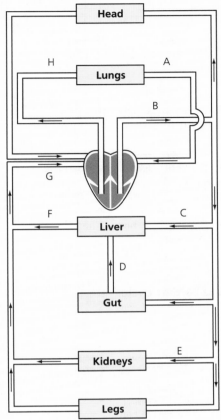

a Name the blood vessels labelled A–H. (AO1) 4 marks

A _____   B _____

C _____   D _____

E _____   F _____

G _____   H _____

b Which part of the body does the coronary artery supply? (AO1) 1 mark

....

....

c The diagram shows a double circulatory system. What is the advantage of
this system? (AO1) 1 mark

....

....

d Which of the blood vessels in the diagram:

i has the lowest pressure? (AO1) 1 mark

....

ii has the highest concentration of amino acids following a meal? (AO1) 1 mark

....

7 Complete the table to describe the pressure changes that affect the movement of valves in the heart. (AO1)

4 marks

Name of valve	Opens or closes?	When
		The pressure in the ventricle is higher than the pressure in the atrium
		The pressure in the aorta is higher than the pressure in the ventricle
		The pressure in the ventricle is higher than the pressure in the aorta
		The pressure in the atrium is higher than the pressure in the ventricle

8 The graph shows the pressure changes in the heart during the cardiac cycle.

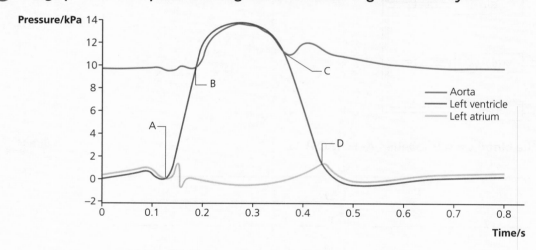

a What is happening at points A, B, C and D? (AO2)

4 marks

A

....

B

....

C

....

D

....

b Between which points is the ventricle emptying? (AO2)

1 mark

....

9 The diagram shows the structure of a capillary, an artery and a vein.

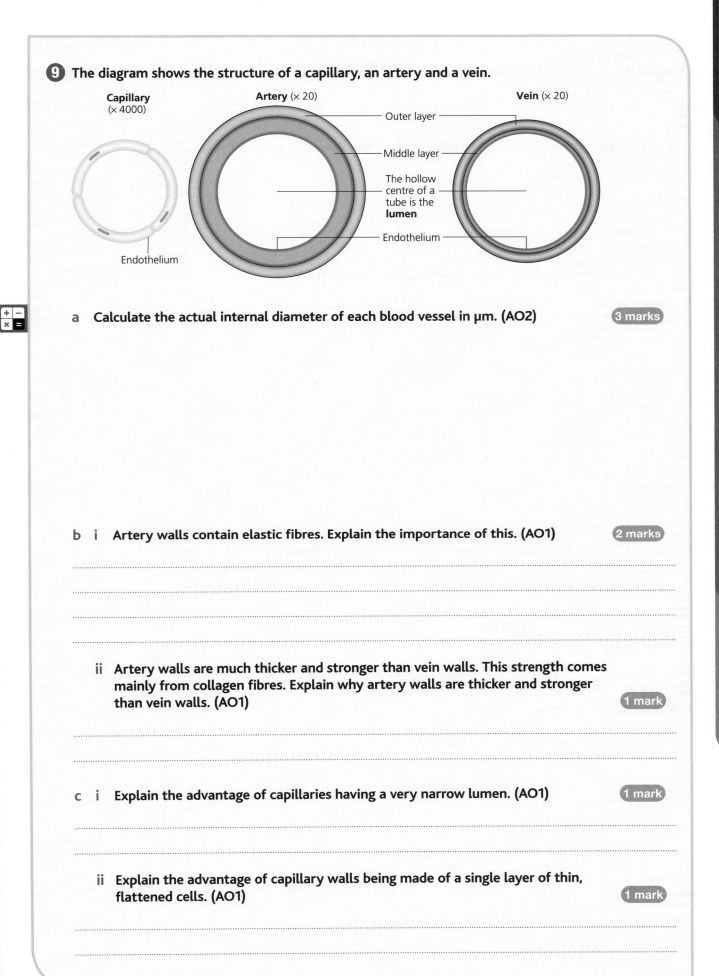

Capillary (× 4000)   Artery (× 20)   Vein (× 20)

Outer layer

Middle layer

The hollow centre of a tube is the **lumen**

Endothelium

Endothelium

a Calculate the actual internal diameter of each blood vessel in μm. (AO2)   3 marks

b i Artery walls contain elastic fibres. Explain the importance of this. (AO1)   2 marks

....

....

....

....

ii Artery walls are much thicker and stronger than vein walls. This strength comes mainly from collagen fibres. Explain why artery walls are thicker and stronger than vein walls. (AO1)   1 mark

....

....

c i Explain the advantage of capillaries having a very narrow lumen. (AO1)   1 mark

....

....

ii Explain the advantage of capillary walls being made of a single layer of thin, flattened cells. (AO1)   1 mark

....

....

10 a How is tissue fluid different from blood plasma? (AO1) `1 mark`

....

b What causes tissue fluid to form at the arteriole end of a capillary? (AO1) `2 marks`

....

....

....

c Describe how tissue fluid returns to the circulation. (AO1) `4 marks`

....

....

....

....

....

....

....

d Filariasis is a condition in which a parasitic worm blocks some of the lymph vessels in the leg. This leads to tissue fluid accumulation in the lower leg (oedema). Explain why. (AO2) `2 marks`

....

....

....

11 Define these terms.

a stroke volume (AO1) `1 mark`

....

....

b cardiac output (AO1) `1 mark`

....

....

....

12 A person has a cardiac output of 6.0 dm³ min⁻¹ and a heart rate of 100 beats per minute. What is her stroke volume? (AO2)

13 A normal resting heart rate is 60 to 80 beats per minute. A trained athlete may have a resting heart rate as low at 40 beats per minute. Explain how. (AO2)

2 marks

....

....

....

....

14 The graph shows mean blood cholesterol levels and death rates from heart disease in men aged 35–74 in several different populations.

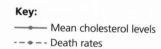

Key:
— Mean cholesterol levels
- - ◆ - - Death rates

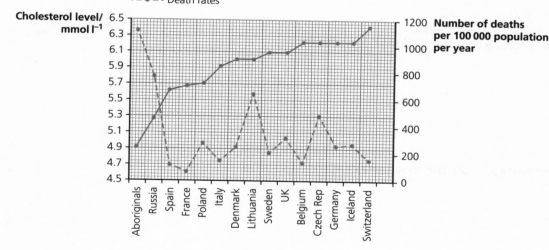

a Death rates are expressed per 100 000 of the population. Explain why. (AO3)

2 marks

....

....

....

b A journalist who saw these data wrote an article entitled 'Cholesterol does not cause heart disease'. Evaluate his conclusion. (AO3)

5 marks

....

....

....

....

....

....

....

....

....

....

....

....

....

Mass transport in plants

The cohesion-tension theory describes how water is transported through the stem and leaves of a plant, mainly via the xylem. The mass flow hypothesis describes how organic substances are transported through the phloem tissue of plants. Transpiration rates can be measured using a variety of practical techniques. There are experimental methods that can be used to investigate the mass flow hypothesis.

1 Describe how:

a water enters a root hair cell (AO1)

2 marks

....

....

....

b water passes across the root cortex (AO1)

2 marks

....

....

....

2 Describe the role of the endodermis in water transport across the root. (AO1)

2 marks

....

....

....

3 How is the structure of xylem adapted to its function of transporting water? (AO1) `4 marks`

....

....

....

....

....

....

....

....

....

....

....

4 Use your knowledge of water potential to explain how water is lost in transpiration from a leaf. (AO1) `2 marks`

....

....

....

....

5 Explain how transpiration draws a column of water up the xylem of a plant. (AO1) `3 marks`

....

....

....

....

....

....

....

....

6 The diagram shows a potometer.

- Leafy twig
- Radius of tube
- Tap
- Markers
- Capillary tube
- Air bubble
- Distance moved by bubble
- Beaker containing water

a How could you use this apparatus to find the volume of water taken up per cm^2 of leaf per minute? (AO3)

4 marks

b How could you use this apparatus to find the effect of air movement on the rate of transpiration? (AO3)

2 marks

7 a What are the following?

i a source (AO1)

1 mark

ii a sink (AO1)

1 mark

b Describe the mass flow hypothesis for the mechanism of translocation of organic substances in plants. (AO1)

6 marks

8 The diagram shows a plant stem that was ringed below the leaves. This means that a ring of bark and phloem tissue was removed. The diagram also shows the appearance of the same stem one week later.

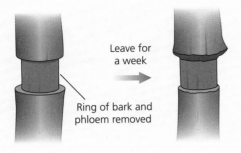

Leave for a week

Ring of bark and phloem removed

a   Explain the appearance of the stem after one week. (AO2)   2 marks

....

....

....

....

b   The xylem is left intact when carrying out a ringing investigation such as this.
     Explain why this is important. (AO2)   2 marks

....

....

....

....

Exam-style questions

1 The llama lives in the Andes mountains where there is a lower partial pressure of oxygen than at sea level. The graph shows the oxyhaemoglobin dissociation curve for llama haemoglobin and for human haemoglobin.

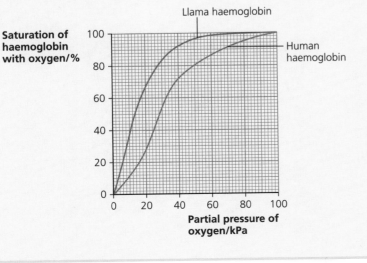

a Use the graph to explain how llamas are better adapted to living at altitude than humans. (AO2)

3 marks

....

....

....

....

....

....

....

....

b i Sketch a line on the graph to show the oxyhaemoglobin dissociation curve you would expect for a fetal llama. (AO2)

1 mark

ii Explain the shape of the line you drew in part i. (AO2)

2 marks

....

....

....

....

2 a Describe *two* ways in which a phloem sieve tube is adapted for its function. (AO1) 2 marks

....

....

....

....

b In an investigation, a scientist supplied carbon dioxide containing radioactive carbon atoms to some of the leaves of a plant. This is shown in Figure 1.

Flower bud

$^{14}CO_2$ supplied

Figure 1

After a few hours, the plant was removed from its pot, the roots washed and the plant placed against photographic film. Figure 2 shows where fogging occurred on the film. Fogging occurs where radioactivity is present.

Figure 2

i The scientists showed that the radioactivity was present in the phloem and not in the xylem. Suggest *one* way in which they could do this. (AO3) `2 marks`

....

....

....

....

ii Give *two* conclusions that you can make from the results of this investigation. (AO2) `2 marks`

....

....

....

3 a The diameter of a tree is very slightly less in the middle of a warm, sunny day than it is in the middle of the night. Explain how this evidence supports the cohesion-tension theory of water movement in plants. (AO2) `4 marks`

....

....

....

....

....

....

....

b One factor that causes water to move up the xylem in a plant is capillarity, i.e. the way that water moves up a small capillary tube for a short distance. The column of xylem vessels in the stem of a plant acts like a capillary tube. Capillary rise is calculated using the following formula:

$$\text{Capillary rise} = \frac{14.9 \times 10^{-6}\ m^2}{r}$$

where r = radius of tube/m

i A typical xylem vessel has a radius of 25 μm. Calculate the distance that water would move up the xylem if capillarity was the only factor involved. Show your working. (AO2)

ii The gaps between the cellulose fibres in the plant cell wall also form tiny capillary tubes that cause water to pass through them. These gaps have a typical diameter of 10^{-8} m. Use this information and the formula above to explain how water evaporating from the surface of the mesophyll cells in a leaf can generate a tension that helps to cause water to move upwards through the xylem. (AO2)

....

....

....

....

Section 4

Genetic information, variation and relationships between organisms

DNA, genes and chromosomes

DNA is the molecule that carries genetic information in all living things. In prokaryotic cells, the DNA molecule is short and circular, and not associated with proteins. In eukaryotes, the DNA is linear and associated with histone proteins to form chromosomes. A gene is a sequence of DNA that codes for RNA and proteins. Three bases in DNA code for one amino acid. In eukaryotes there is a lot of non-coding DNA, both between genes and within them.

1 Complete the table with the words that match the definitions. (AO1) **6 marks**

Word	Definition
	Three bases in DNA that code for one amino acid
	Fixed position on a chromosome where a gene is found
	A sequence of DNA that codes for functional RNA and/or the amino acid sequence of a polypeptide
	Non-coding DNA found within the genes of eukaryotes
	Sequence of DNA within a gene that codes for a protein
	Protein molecules that associate with the DNA in a eukaryotic chromosome

2 Chloroplasts and mitochondria contain DNA. How does this DNA resemble that of prokaryotes? (AO1) **1 mark**

....

....

DNA and protein synthesis

The genome consists of all the genes in a cell and the proteome is all the proteins that a cell can produce. mRNA and tRNA have specific structures that enable them to carry out their function of making proteins.

Protein synthesis occurs in two stages. Transcription is the production of mRNA from a DNA template and translation is the synthesis of proteins using mRNA, ATP and tRNA.

1 Give *one* way in which the structures of mRNA and tRNA are:

a   alike (AO1)

1 mark

....

....

b   different (AO1)

1 mark

....

....

2 Arrange these sentences in the right order to describe the stages of protein synthesis. (AO1)

2 marks

A	This continues until a non-sense or 'stop' codon is reached and the polypeptide is released
B	RNA nucleotides align alongside the exposed DNA bases by complementary base pairing
C	A tRNA molecule brings its specific amino acid to the ribosome and its anticodon binds to the first mRNA codon by complementary base pairing
D	The introns are spliced out and the exons are combined to make functional mRNA
E	A section of the DNA in the nucleus unzips and one strand becomes a template
F	A second tRNA with the appropriate anticodon enters the ribosome, bringing its amino acid
G	The enzyme RNA polymerase joins the RNA nucleotides together to make a molecule of pre-mRNA
H	The first tRNA leaves the ribosome and the ribosome moves along the mRNA
I	The mRNA leaves the nucleus via a nuclear pore and attaches to a ribosome
J	The amino acids join together by a peptide bond

....

3 Distinguish between:

a   the proteome and the genome (AO1)

2 marks

....

....

....

....

b   a codon and an anticodon (AO1)

2 marks

....

....

....

....

c   introns and exons (AO1)

2 marks

....

....

....

....

4 Complete the table to show the mRNA base sequence and tRNA anticodons that the given DNA base sequence codes for. (AO1)

2 marks

DNA base sequence	A T T G C C A G C T G A
mRNA base sequence	
tRNA anticodons	

5 DNA is said to be universal and non-overlapping. Explain what is meant by these terms.

a   universal (AO1)

1 mark

....

....

b   non-overlapping (AO1)

1 mark

....

....

6 What is the role of ATP in protein synthesis? (AO1)

1 mark

....

....

....

....

Genetic diversity can arise as a result of mutation or during meiosis

A gene mutation is a change in the base sequence of the DNA. Examples of gene mutation include deletion and substitution. Although mutations occur spontaneously, mutagenic agents can increase the rate of mutation. Chromosome mutations may also occur during meiosis, resulting in daughter cells with additional or missing chromosomes. Meiosis is a form of nuclear division that results in four daughter cells that are haploid and genetically different from each other. Variation in meiosis occurs as a result of independent segregation and crossing-over.

1 The diagram shows a section of DNA before replication and two possible outcomes after replication.

Outcome A   Outcome B

a   Name the type of mutation that has occurred in each outcome.

   i   outcome A (AO1)   1 mark

   ....

   ii   outcome B (AO1)   1 mark

   ....

b   Explain *two* possible results of the mutation shown in outcome A. (AO1)   6 marks

....

....

....

....

....

....

....

....

....

....

c   Explain the possible result of the mutation shown in outcome B. (AO1)   `3 marks`

....

....

....

....

....

2 The DNA code is said to be degenerate. What does this mean? (AO1)   `2 marks`

....

....

....

....

3 What is meant by a 'mutagenic agent'? Give *two* examples. (AO1)   `2 marks`

....

....

....

4 Complete the table to show the mass of DNA and the number of chromosomes in cells at different stages of meiosis. (AO2)   `2 marks`

Stage	Number of chromosomes per cell	Mass of DNA per cell/ arbitrary units
Immediately before meiosis	28	40
End of the first division of meiosis		
End of the second division of meiosis		

5 What are homologous chromosomes? (AO1)   `2 marks`

....

....

....

....

6 Explain what crossing over is and how it results in genetic variation. (AO1)   `3 marks`

....

....

....

....

....

....

....

....

31

7 Explain how independent segregation results in genetic variation. (AO1) `3 marks`

....

....

....

....

....

....

8 What is non-disjunction? (AO1) `3 marks`

....

....

....

....

....

....

9 How does random fertilisation increase genetic variation within a species? (AO1) `2 marks`

....

....

....

....

10 Give *two* differences between mitosis and meiosis. (AO1) `2 marks`

....

....

....

11 The diagram shows the life cycle of a fungus. (AO2) `2 marks`

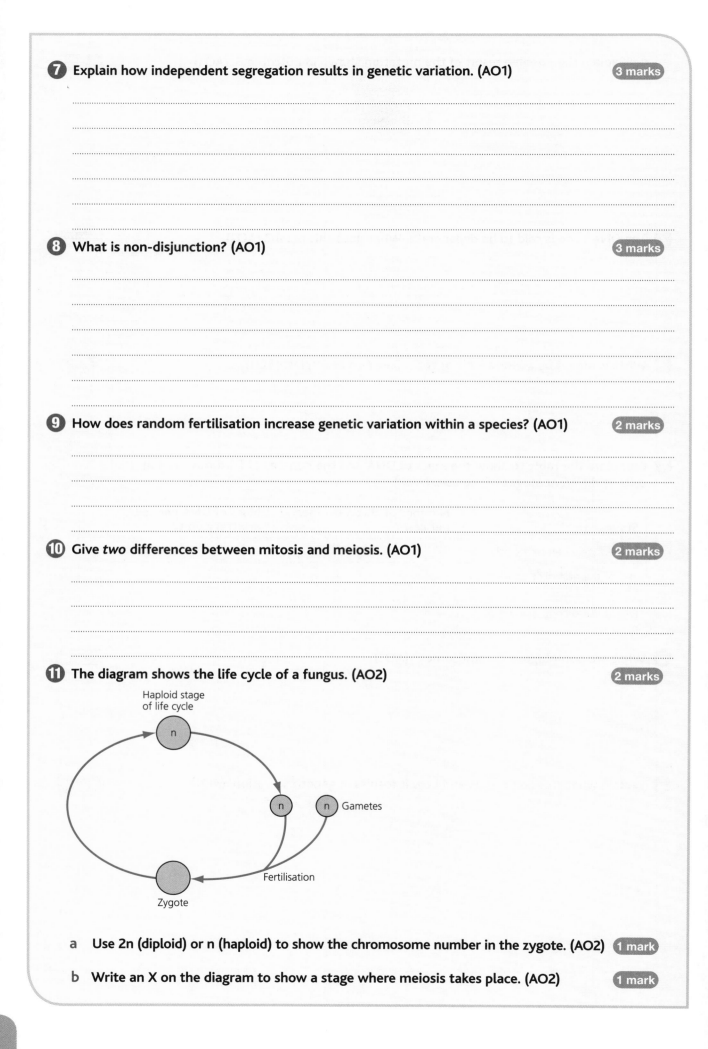

Haploid stage of life cycle

n

n   n Gametes

Fertilisation

Zygote

   a   Use 2n (diploid) or n (haploid) to show the chromosome number in the zygote. (AO2) `1 mark`

   b   Write an X on the diagram to show a stage where meiosis takes place. (AO2) `1 mark`

Genetic diversity and adaptation

Genetic diversity is the number of different alleles in a population. This enables natural selection to occur. New alleles arise by random mutation. Although most mutations are harmful, a new allele may occasionally be beneficial and confer an advantage to its possessor. This individual will be more likely to survive, reproduce and pass on the allele to its offspring. Two kinds of natural selection are stabilising selection and directional selection. Natural selection results in organisms that are better adapted to their environment, including anatomical, physiological or behavioural adaptations.

---

1 The spiny cactus lives in desert conditions. Spines on the plant are genetically determined. Peccaries are mammals that live in the desert. They eat spiny cacti, especially those with the fewest spines. Parasitic insects also live in the desert. They lay their eggs in the cacti, gradually killing the plant. The parasitic insects tend to infect those cacti with the densest spines.

   a   What kind of natural selection is likely to be acting on the spiny cactus? Explain your answer. (AO2)   **3 marks**

   ....

   ....

   ....

   ....

   ....

   ....

   b   Cacti have a special kind of photosynthesis that means they can take in carbon dioxide at night and store it for use in the daytime. Explain the advantage of this. (AO2)   **2 marks**

   ....

   ....

   ....

   ....

2 Warfarin is a poison used to kill rats. Rats that are resistant to warfarin need a lot of vitamin K in their diet. Warfarin resistance can occur as a result of a chance mutation. In the 1960s warfarin was used regularly as rat poison. A few years later, there were populations of rats that were resistant to warfarin.

   a   Explain how natural selection has led to these populations of warfarin-resistant rats. (AO2)   **4 marks**

   ....

   ....

   ....

   ....

   ....

   ....

b   What type of natural selection has taken place? (AO1)

1 mark

....

3   Explain why:

a   human birth mass is an example of stabilising selection (AO1)

2 marks

....
....
....
....

b   the development of antibiotic resistance in bacteria is an example of directional
    selection (AO1)

4 marks

....
....
....
....
....
....
....
....
....

Exam-style questions

1. African elephants, *Loxodonta africana,* are often killed for their tusks. In 1990, 10.5% of the elephants in a national park in Zambia were tuskless. By 2010, 38.2% of the elephants in the national park were tuskless.

   a i   What kind of selection is shown in this example? (AO1)   1 mark

   ....

      ii Use your knowledge of natural selection to explain this change. (AO2)   4 marks

   ....
   ....
   ....
   ....
   ....
   ....
   ....
   ....
   ....
   ....
   ....
   ....
   ....

   b Both the African and the Asian elephant are believed to have evolved from mammoths.

   - Based on mitochondrial DNA studies, mammoths and Asian elephants are more closely related to each other than either of them are related to African elephants.

   - Studies of physical characteristics show that mammoths are more closely related to Asian elephants than to African elephants.

   - Some studies of chromosomal DNA suggest that mammoths are more closely related to African elephants, whereas other chromosomal DNA studies suggest that mammoths are related equally to Asian and African elephants.

   These studies give conflicting information. Suggest *two* reasons for this. (AO3)   2 marks

   ....
   ....
   ....
   ....
   ....
   ....
   ....

2 To obtain a karyotype, cells from an individual are cultured in a growth medium and a spindle inhibitor is added. The karyotype is made by photographing the chromosomes in a cell and arranging them in homologous pairs. The diagram shows the karyotype of an individual.

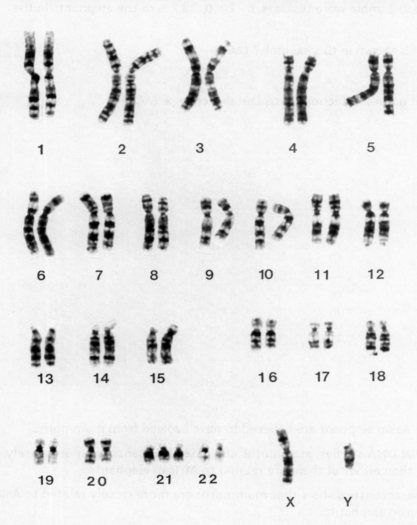

a i The chromosomes are arranged in homologous pairs. What are homologous chromosomes? (AO1)

2 marks

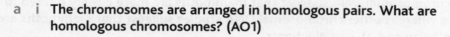

....

....

....

ii At what stage in the cell cycle were the chromosomes photographed? Explain your answer. (AO2)

3 marks

....

....

....

....

....

b The diagram indicates that a mutation has occurred in this individual. Identify this mutation and explain how it has occurred. (AO1)

3 marks

....

....

....

....

....

Species and taxonomy

A species is a group of organisms with observable similarities that can interbreed to produce fertile offspring. In some species, courtship behaviour is necessary before successful mating can occur. Species can be classified into groups, called taxa, based on their evolutionary origins. Each species has a binomial name consisting of its genus and species. Organisms are placed in their different taxa using evidence such as immunology and DNA sequencing.

1 Organisms are classified into a phylogenetic hierarchy. Explain what is meant by these terms.

a   phylogenetic (AO1)   `1 mark`

....

....

b   hierarchy (AO1)   `1 mark`

....

....

2 Give *two* reasons why courtship behaviour is necessary before mating can occur in many species. (AO1)   `2 marks`

....

....

....

....

3 a   Complete the table to show the classification of the emperor penguin, *Aptenodytes forsteri.* (AO1)   `2 marks`

Domain	Eukaryota
	Animalia
	Chordata
	Aves
	Sphenisciformes
	Spheniscidae
Genus	
Species	

b   Two other kinds of penguin are the king penguin, *Aptenodytes patagonicus,* and the Humboldt penguin, *Spheniscus humboldti.* What do their binomial names tell you about the relationship between these three different species of penguin? (AO1)   `2 marks`

....

....

....

....

④ Some scientists wanted to find the relationship between four different species of snake. They injected some serum from species A into a rat. The rat produced antibodies against species A serum. The scientists then put serum from snake species A, B, C and D into four separate tubes. They added the rat antibodies to each tube. The table records the amount of precipitate in each tube.

Type of serum in tube	Amount of precipitate formed with rat antibodies against species A
Species A	+++++
Species B	++++
Species C	+
Species D	++

What do these data indicate about the relationship between species A and the three other species of snake? Explain your answer. (AO2)   **2 marks**

....

....

....

⑤ How can genome sequences be used to compare relationships between species? (AO1)   **2 marks**

....

....

....

....

Biodiversity within a community

It is important that we consider biodiversity in all kinds of habitats, from very small areas to the whole of the Earth. An index of diversity measures the number of species in a community as well as the number of individuals of each species. Species richness is a measure of the number of different species in a community. One factor that reduces biodiversity is farming. There needs to be a balance between farming and the need for conservation.

① Some scientists studied an area of land and recorded the following data.

Species	Number of plants
Plantain	104
Creeping buttercup	42
Hawkbit	22
Speedwell	17
Daisy	15
Nettle	9
Dandelion	7

a   Calculate the species richness of this area. (AO1) 1 mark

....

b   Calculate the index of diversity for this area using the following formula. (AO2)   2 marks

$$d = \frac{N(N-1)}{\sum n(n-1)}$$

c   Evaluate the use of an index of diversity to measure the biodiversity of
an area. (AO2)   2 marks

....

....

....

2 Describe *four* ways in which farming may reduce biodiversity. (AO1)   4 marks

....

....

....

....

....

....

....

....

3 The following are conservation measures that conservation organisations encourage farmers to
carry out. Suggest how each of these measures may improve biodiversity.

a   Leaving the field margins (areas around the edges of fields) uncultivated and free
of pesticides. (AO2)   3 marks

....

....

....

....

....

b   Leaving stubble (the roots and lower parts of a crop) in the field over winter after
the crop has been harvested. (AO2)   1 mark

....

....

4 The graph shows the development of resistance to a particular fungicide in a specific pest fungus.

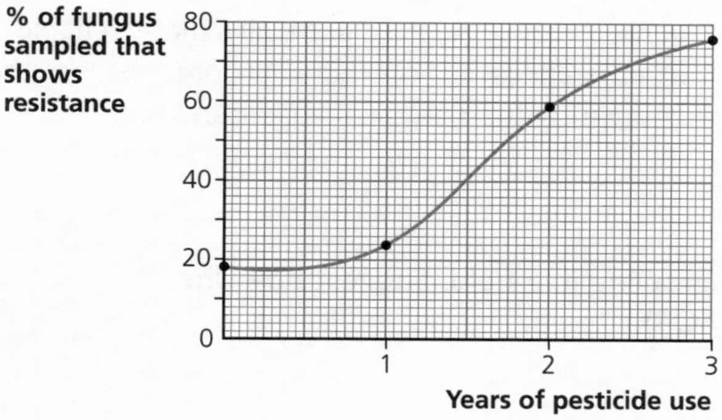

% of fungus sampled that shows resistance

Years of pesticide use

a   Describe the graph. (AO2)

2 marks

....

....

....

b   Explain the graph. (AO2)

4 marks

....

....

....

....

....

....

....

....

Investigating diversity

There are various ways of measuring genetic diversity within or between species. These include measuring observable characteristics, comparing the base sequences of DNA or mRNA, or comparing the amino acid sequence of proteins. The more alike any two organisms are when these comparisons are made, the more closely related they are. With advances in gene technology, it is generally more reliable to look at differences in DNA between organisms rather than comparing their physical appearances.

1 A scientist wanted to sample dandelion plants growing in a field. Suggest how she could obtain a random sample of dandelions. (AO3) `4 marks`

2 The table shows the amino acid sequence found in a section of human cytochrome C and in the corresponding section of cytochrome C in some other organisms.

Organism	Amino acid sequence											
	1	2	3	3	5	6	7	8	9	10	11	12
Human	Gly	Asp	Val	Glu	Lys	Gly	Lys	Lys	Ile	Phe	Ile	Met
Pig	Gly	Asp	Val	Glu	Lys	Gly	Lys	Lys	Ile	Phe	Val	Gln
Chicken	Gly	Asp	Ile	Glu	Lys	Gly	Lys	Lys	Ile	Phe	Val	Gln
Dogfish	Gly	Asp	Val	Glu	Lys	Gly	Lys	Lys	Val	Phe	Val	Gln

Based on these data only, which organism is most closely related to humans? Explain your answer. (AO2) `2 marks`

3 DNA hybridisation can be used to compare the DNA sequences of different organisms. In summary, double-stranded DNA is taken from species A. Then hybrid DNA is formed containing one strand of DNA from species A and one strand from species B. The temperature required for 50% of the DNA sample to become single-stranded is recorded.

a i Why is heat needed to make the DNA samples single-stranded? (AO2)   1 mark

....

ii Double-stranded DNA from species A becomes single-stranded at a higher temperature than the hybrid DNA. Explain why. (AO2)   2 marks

....

....

....

....

b The difference between the temperature needed to make 50% of the double-stranded species A DNA single-stranded, and that needed for the hybrid species A/species B DNA is called $T_{50}H$. The table shows the $T_{50}H$ values for some different organisms.

Organisms being compared	$T_{50}H$
Common chimp and pygmy chimp	0.73
Common chimp and gorilla	2.30
Common chimp and human	1.68

The cladogram below shows the relationships between some species that are believed to have shared a common ancestor 25 million years ago.

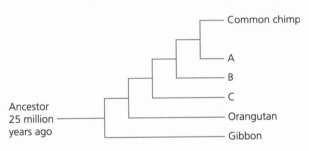

Use the data above to identify species A, B and C in the cladogram.

i species A (AO2)   1 mark

....

ii species B (AO2)   1 mark

....

iii species C (AO2) 1 mark

....

4 Examining DNA rather than comparing the physical appearances of organisms to determine their genetic similarities is usually more reliable. Explain why. (AO1)   `2 marks`

....

....

....

....

5 The table records the mean length of dandelion leaves in two different areas and the standard deviation of each sample.

Area	Mean leaf length/cm
Roadside verge	16.2 ± 4.3
Football field	13.1 ± 2.1

a   What does standard deviation indicate? (AO2)   `1 mark`

....

....

b   Why is standard deviation used, rather than the range? (AO2)   `1 mark`

....

....

c   A student concluded that dandelion leaves in the football field have shorter leaves than those on the roadside verge. Is this a reliable conclusion? Explain your answer. (AO2)   `2 marks`

....

....

....

....

Exam-style questions

1 Complete the table to show the classification of the African elephant, *Loxodonta africana*. (AO1)   **2 marks**

	Eukaryota
	Animalia
	Chordata
	Mammalia
	Proboscidae
	Elephantidae
Genus	
Species	

2 A scientist injected some human serum into a rabbit. He did this on three occasions at monthly intervals. After this, he took a blood sample from the rabbit and purified some anti-human antibodies from it.

a The scientist injected the serum into the rabbit three times at monthly intervals. Explain why. (AO2)   **2 marks**

....

....

....

....

b The scientist placed equal volumes of serum from various animals in separate test tubes. He added an equal volume of anti-human antibodies to each tube. After 30 minutes he measured the amount of precipitate formed. This was measured relative to the human sample, which was taken to be 100%. The results are shown in the table.

Organism	Amount of precipitate formed/%
Human	100
Chimpanzee	97
Gorilla	92
Gibbon	79
Baboon	75
Spider monkey	58
Lemur	37
Hedgehog	17
Pig	8

i The anti-human antibody produced much more precipitate when mixed with gorilla serum than when it was mixed with spider monkey serum. Explain why. (AO2) `3 marks`

....

....

....

....

....

ii Data like these are important when classifying organisms. Explain how. (AO2) `3 marks`

....

....

....

....

....

3 Scientists measured the species richness of organic farms and conventional farms. They calculated the % difference in species richness in organic farms relative to conventional farms. Their results are shown in the table.

Group of organisms	Difference in species richness/%
Plants	75
Birds	21
Arthropods	20
Microbes	19
Mean for all organisms	**27**

a **i** What is species richness? (AO1) `1 mark`

....

ii How would the % difference in species richness have been calculated? (AO2) `1 mark`

....

....

b Give *two* conclusions that can be drawn from these data. (AO3) `2 marks`

....

....

....

....

c Organic farms do not use chemical pesticides. Suggest how this might increase biodiversity on farmland. (AO2)

3 marks

....

....

....

....

....

....

Also available

...and many more

Go to http://www.hoddereducation.co.uk/studentworkbooks for details of all our student workbooks.

Philip Allan, an imprint of Hodder Education, an Hachette UK company, Blenheim Court, George Street, Banbury, Oxfordshire OX16 5BH

Orders
Bookpoint Ltd, 130 Milton Park, Abingdon, Oxfordshire OX14 4SB
tel: 01235 827827
fax: 01235 400401
e-mail: education@bookpoint.co.uk
Lines are open 9.00 a.m.–5.00 p.m., Monday to Saturday, with a 24-hour message answering service. You can also order through www.hoddereducation.co.uk

© Pauline Lowrie 2015
ISBN 978-1-4718-4465-2
First printed 2015
Impression number   5   4
Year   2020   2019   2018   2017

Cover photo reproduced by permission of Fotolia

Photo p. 36 © L. Willat, East Anglian Regional Genetics Service/SPL

Printed in Dubai

Hachette UK's policy is to use papers that are natural, renewable and recyclable products and made from wood grown in sustainable forests. The logging and manufacturing processes are expected to conform to the environmental regulations of the country of origin.

ISBN 978-1-4718-4465-2

9 781471 844652